THE IRON ROAD

THE IRON ROAD

A Portrait of American Railroading

Photographs by David Plowden • Text by Richard Snow

Four Winds Press New York

LIBRARY OF CONGRESS CATALOGING IN PUBLICATION DATA

Snow, Richard F.
The iron road.

SUMMARY: A history of American railroading with a commentary on its status today.
1.Railroads — United States — Juvenile literature.
[1.Railroads — History] I.Plowden, David.
II.Title.
TF23.S66 385'.0973 78 – 5388
ISBN 0-590-07523-3

Published by Four Winds Press
A division of Scholastic Magazines, Inc., New York, N.Y.

Printed in the United States of America
Library of Congress Catalog Card Number: 78 – 5388
1 2 3 4 5 82 81 80 79 78

For all the railroad men who built the iron road and who run it today.

WHEN I WAS a small boy in the early 1950s, my father sometimes drove me over to Yonkers to watch the trains going by on the Putnam Division. The Putnam Division was a quiet little spur of the mighty New York Central; it peeled off from the main line in Manhattan and ran through a couple of dozen Westchester towns before it hooked up again at Brewster, fifty miles to the north. Not much happened on the Putnam Division. My father and I would walk along the tracks for a mile or so, maybe find a railroad spike to take home, and I would pick my way along on the top of a rail, keeping my balance as long as I could. Once in a great while, a train went by—two or three prehistoric day coaches pulled by a steam locomotive. My father and I would wave and, if the engineer was in a good mood, he'd respond with a couple of hot, brassy blasts on the whistle.

Being a kid, I would have been happier spending my day watching people shoot each other in a movie. But after the occasional train had gone by and the engineer had favored us with a tug on his whistle cord, I would stand watching the tracks curving into

Colorado & Southern Railway near Clayton, New Mexico.

the woods, listening to the diminishing clatter, content in my child's knowledge that something rather wonderful had happened. And I was right.

If I'd had the strength to match my enthusiasm, I could have walked that rail up to Brewster and then teetered along north by west to Chicago—or, if I chose, south by west to St. Louis—then straight west across cornfields and wheat fields and mountains until, never once having stepped off the tracks, I could see the Pacific Ocean combing in from Asia. Like every other small railroad line, the Putnam Division fed the great lines. The whole, huge country was—and is—laced together with steel rails.

A lot of people have stood as I stood, looking down the tracks. One of the first was a man named Charles Carroll, who, unlike the rest of us, looked down tracks that weren't yet there. He was a staid, sour old man who had lived a very long time. In the stifling Philadelphia summer of 1776, he had waited his turn with some other men and then had picked up a pen and put his name on the Declaration of Independence. He must have been a man willing to take a gamble because he and the men who signed with him faced even odds of being hanged as traitors.

His gamble paid off, and he became one of the Founding Fathers, rich and respected. By 1827 all the other men who had signed the document were dead and gone. But old Carroll was able to stand holding a shovel on a hot July Fourth while a respectful

Semaphore signal bridges, Jersey City, New Jersey.

crowd waited for him to break the ground for America's first real railroad — the Baltimore & Ohio. Carroll poked his shovel at the proper place, scraped up some dirt, and then stood back while people cheered. A band played and cannon went off, but it couldn't have been much of a ceremony, because the B & O Railroad had no track, no cars, and no engines. And yet Carroll said, "I consider this among the most important acts of my life, second only to my signing the Declaration of Independence, if even it be second to that."

The machine that would justify Carroll's vision had been born, breathing fire and obscure promise, more than a century before in the mines of England. Chief among the difficulties miners faced was the water that constantly seeped into their diggings. An army officer named Captain Thomas Savery tackled the problem and developed a pump based on the ancient observations that boiling water expands violently into steam and that steam, condensing back into water, creates a vacuum. Savery's engine was slow, ponderous, big as a house; steam entered a cylinder where a spray of cold water condensed it. The resulting vacuum sucked water up out of the mine, a new jet of steam drove it out of the cylinder, and the laborious process began again.

A blacksmith named Newcomen added a piston to Savery's engine and, in the 1760s, James Watt improved the whole business immensely by moving the condensing process to a second chamber. But the machine was still big, clumsy, weak, and absolutely rooted to one spot.

And so things stood until the brother of a young Delaware wagonmaker named

Freight yard, Bluefield, West Virginia.

Oliver Evans filled a gun barrel with water, stopped up the ends, and tossed it into a fire. When the water turned to steam, the barrel exploded. His brother was delighted merely by the satisfactory bang, but Evans saw in that explosion a force that could change the history of nations, and the conviction never left him. He began to study, and when he came across an account of Newcomen's engines, he said, "But he's doing it the wrong way!"

The mistake, Evans realized, lay in the reliance on the vacuum; he developed an engine with sliding valves that let the steam punch the piston on both strokes instead of just one. With that done, the engine could be made as small as anyone wanted; all it required was high-pressure steam.

Evans built his engines but suffered the usual inventor's lot of being mocked. He died in 1819, a thoroughly unhappy man, leaving behind his most absurd proposal: "The time will come when people will travel in stages moved by steam engines, from one city to another, almost as fast as birds fly, fifteen to twenty miles an hour. Passing through the air with such velocity, changing the scene in such rapid succession, will be the most exhilarating exercise. A carriage (steam) will set out from Washington in the morning, the passengers will breakfast at Baltimore, dine at Philadelphia, and sup in New York on the same day."

Yet for all Evans's certainty, it was England and not America that pioneered the railroad. In 1784 an assistant of Watt's named Murdock built a tiny, self-propelled steam carriage and, fearing ridicule, turned it loose at night. The little engine sped along perfectly, but Murdock promptly lost interest in it; his only practical accomplishment had been to terrify a clergyman who, seeing the machine streak past, burning and fizzing, was convinced he had met Satan.

Still, Murdock's creation was not wasted, for it caught the attention of an eccentric, evil-tempered inventor named Richard Trevethick who, twenty years later, completed the first real railroad engine. In 1804 this fantastic collection of cogs and gears pulled five cars over the rails in a Welsh iron foundry. It ran only once — the heavy engine chewed up the track — but the railroad had been born.

Able men followed Trevethick's lead. In 1825, the twelve-mile-long Stockton and Darlington Railway opened amid great celebration, and five years later the Liverpool & Manchester Railway, an awesome piece of engineering, began regular operations.

So England got there first, and thereafter the English tended to be a bit condescending toward American railroading efforts. But English railroads served a very different purpose than the American ones would. In small, settled, civilized England, with all the cities built and the roads connecting them traveled for centuries, the trains did nothing more than move freight and people from one place to another more quickly than had been possible before. But in America, where a growing population was beginning to spill westward from the Atlantic Coast into a vast and unknown interior, the railroad would have a much bigger job to do. It would build, shape, and define the country, would summon cities out of forests, would give us strong legends and a brand-new culture; ultimately, it would make us a world power.

If old Carroll had vision enough to see all this, most other Americans did not. Man had never before been able to travel faster than the fastest horse could run, and a lot of people felt that God meant things to stay that way. By the time the idea of railroads started to stir over here, Americans interested in travel had put a lot of their money into building canals. These waterways could move freight and people at about three miles per hour, and had to be shut down during the winter months. Nevertheless, the men who had backed them attacked the railroads, just the way frightened people will attack any big, new, half-understood idea. One canal stockholder spoke with particular eloquence: "I see what will be the effect of it; it will set the whole world a-gadding. Twenty miles an hour, sir! Why, you will not be able to keep an apprentice boy at his work; every Saturday evening he must take a trip to Ohio to spend the Sabbath with his sweetheart. Grave, plodding citizens will be flying about like comets. . . . And then, sir, there will be barrels of pork and cargoes of flour, and chaldrons of coal, and even lead and whiskey and such-like sober things, that have always been used to sober traveling, whisking away like a set of sky-rockets. It will upset the gravity of the nation."

But the railroad was coming, and the canal men couldn't stop it. In 1830, while Carroll's B & O was still experimenting with trains pulled by horses and pushed by wind —and with an overrated toy steam engine called the *Tom Thumb*—the Charleston & Hamburg Railroad had the first true American-built locomotive hammered together in a New York foundry. Christened "The Best Friend of Charleston," it pulled a train of

Westbound freight near Havre, Montana.

cars over six miles of track that December, thereby becoming the first steam engine to do so in the United States. Half a year later, it had the doleful honor of being the first locomotive to explode. It blew up because the fireman, irritated by the hissing of the safety valve, decided to tie down the valve lever. After a few moments of soothing silence, the boiler burst, killing the fireman and scalding the engineer.

It seems an incredible error on the fireman's part, but in those primeval days, nobody really knew much about railroading. The nation had to learn to use this new thing by trial and error—and nothing people had known before seemed to be of much help. For instance, many argued that a railroad should be open to anybody who wanted to drive a train along it and could pay the toll; that, after all, was the way a turnpike worked. And there was the problem of cows on the tracks, which was tackled in a grisly way by the first cowcatcher, a pronged attachment on the front of the locomotive which, rather than nudging the animal out of the way, speared it.

The first rails were made of wood, with flat iron bars nailed along the top. Under the train's weight, the bars often came loose and curled up from the rail; trainmen gave them the vivid name of "snakeheads." Snakeheads could shoot up through the floor of a car and slice through a passenger. On one train a snakehead came entirely clear, pierced a woman's thigh and went on up through the roof of the car, leaving her transfixed like a carousel animal on a vertical shaft. It took men working with saws and hammers an hour to free her. She lived, but many others died before wrought-iron rails came into use.

Wherever the rails went during the 1830s, there were collisions, explosions, derailments, as the amateurs learned their jobs in a din of small catastrophes. Alarmed people fought the machine. "MOTHERS LOOK OUT FOR YOUR CHILDREN!" one poster shrilled from the walls and fenceposts in Philadelphia. "ARTISANS, MECHANICS, CITIZENS! When you leave your family in health, must you be hurried home to mourn a DREADFUL CASUALTY! . . . Regardless of your interests, or the LIVES OF YOUR LITTLE ONES, THE CAMDEN AND AMBOY are laying a LOCOMOTIVE RAIL ROAD! . . . RALLY PEOPLE in the Majesty of your Strength and forbid this OUTRAGE!"

Despite such assaults, the railroad men stuck by their fledgling locomotives, and in time learned how to handle them. At first, many of the new lines ordered engines built in England, but they soon found them to be all wrong for American roads. Heavy, chubby, stiff, and built for the straight, fast British rights-of-way, they mauled American rails and tended to jump the tracks on tight turns. The American builders modified the British design and put four small wheels in front of the big driving wheels. The engines began to take on the peculiarly American look that would be with them throughout the age of steam: lean and rangy, with the rods that moved the wheels on the outside, where they were easy to get at, rather than inside, where they looked more tidy.

The cars the engines pulled began to change, too. At first they had been nothing more than stagecoaches, fat at the top, narrow at the bottom. But these top-heavy boxes tipped over at the slightest jolt. The railroaders replaced them with lower, longer, sturdier cars and began to put in some comfortable seats. The seats came as a surprise to a lot of people who had thought that trains would only be used to carry freight. Now that there were railroads, however, folks seemed to be moving around more, and you couldn't ask someone who had paid good money to bump along on top of a carload of barrels or a pile of shingles.

By 1840 there were some three thousand miles of track in the country, and the railroaders were winning their last skirmishes with their opponents. The final contests were not against the canal men, but the clergy. Religion held an enormous sway over America at that time and, while many men of God were willing to concede that the steam locomotive might not be a work of the devil, they were certainly not willing to let the trains run on Sundays. By now the railroads had made themselves felt in a hundred ways, but if you had to choose the one day they really came of age, you could do worse than to pick a Sunday morning at the station in Galesburg, Illinois. The train had come in and was loading up when a stern Christian gentleman walked over to the engineer and told him to take his locomotive back to the roundhouse. Twenty years before, even five years before, the engineer would have listened to the voice of authority. But now he was in charge of a lot of hot, expensive metal, and he had a schedule to meet.

"Who are you to give me such orders?" he asked.

Passengers boarding the *Phoebe Snow,* Scranton, Pennsylvania.

EWS STAND

Eastern Standard Time

John M. Caffrey, conductor, Lehigh Valley Railroad.

(right) Pennsylvania Station, New York, New York.

"I am President Blanchard of Knox College, and again I order you to take that engine to the roundhouse, and not to run this train on Sunday."

The engineer thought it over. "Well, President Blanchard of Knox College," he said, "you can go to hell and mind your own business, and I'll take my train out as ordered."

He did, and that was that. Some roads eventually compromised with the church by having the conductors come through the cars and read scripture to Sunday passengers, but the railroad was clearly starting to change the habits of the whole country.

Henry David Thoreau, who brooded about all things mechanical, paid special attention to the trains that passed by within earshot of Walden Pond. "They come and go," he wrote, "with such regularity and precision, and their whistle can be heard so far, that the farmers set their clocks by them, and thus one well-regulated institution regulates a whole country. Have not men improved somewhat in punctuality since the railroad was invented? Do they not talk and think faster in the depot than they did in the stage-office?"

In fact, the railroads were not nearly so precise and reliable as Thoreau made them sound. During the 1840s and 1850s trains often operated erratically on highly flexible schedules. Moreover, there seemed to be a devil-may-care attitude on the part of the railroad men that appalled at least one passenger. He was a Briton named Charles Richard Weld, who, like many Englishmen of the era, had come to America to have a look around and write a book about his impressions. On his way from Cincinnati to Washington, he got aboard a train which started out hours late. The conductors came by and told Weld that they could make up the time, but it would take "smart work." The smart work turned out to be a rate of speed that terrified the Englishman. "The conductor, however, was a determined man and as he evidently attached little value to his own life it was not to be expected that his passengers would be much cared for."

The train rocketed along the curving banks of the Potomac, and Weld, fearing for his life, got a "significant hint of the impending catastrophe . . . by the fall of a ponderous lamp glass on my head. . . ." Another lamp soon flew into the lap of a woman who was sitting nearby, and a passenger, coolly stating that he was an old hand when it came to railroad accidents, suggested that Weld hang on to the seat in front of him, and swivel sideways so his knees wouldn't take the full shock when the accident came.

Sure enough, six miles outside of Harpers Ferry the train went crashing off the tracks and Weld's car flipped over. "I have no distinct recollection," he said, "how I crawled out of the car, for I was half stunned, but I remember being highly delighted when I found my limbs sound. On looking around, the spectacle was extraordinary. With the exception of about half the middle car and engine there was scarcely a portion of the train that was not more or less broken."

Furious, Weld proposed to the other passengers that they complain about the conductor's recklessness to the directors of the line. To his astonishment, they told him

2408

to shut up and mind his own business. After all, the conductor was only doing his job by trying to get them into Harpers Ferry on time. "Accidents on railways," Weld concluded gloomily, "are thought so little of in America it is useless to remonstrate."

Weld was baffled by the whole thing, but another English visitor, a Mrs. Houstoun, came pretty close to explaining the calm acceptance of near disaster when she wrote, "I really think there must be some natural affinity between Yankee 'keep-moving' nature and a locomotive engine. . . . Whatever the cause, it is certain that the humans seem to treat the 'ingine,' as they call it, more like a familiar friend than as the dangerous and desperate thing it really is."

Somewhere along the way, Americans had ceased to think of the railroad as a possibly useful curiosity, and had begun to love it. One of the strangest and truest indicators of this new affection can be found on the tombstones of people killed in train accidents. Despite their grief, the relatives who ordered the monuments sometimes felt such pride in the way their loved ones died that they ordered a locomotive or a passenger car carved on the gravestone.

This same pride could be seen in the locomotives themselves, for the builders—forty of them by 1850—painted their creations like circus wagons. A cab might be bright green, picked out with yellow pinstriping, and decorated with a painting of dawn breaking over a mountain range. The headlamp would have another painting—an Indian in a war bonnet, perhaps—and the sky-blue tender would bear a painted scroll

Steam locomotives, Canadian Pacific Railway.

6036
ENOLA
EMERGENCY
FUEL CUT-OUT

from which gold leaf flashed out the name of the engine—*Ajax*, or *Phoenix*, or *General Scott*.

The railroads assigned each of these ornate locomotives to a single engineer, who would watch over it with the ferocious attention a captain gave his ship. One old railroad man recalled hanging around the roundhouse as a boy, hoping in vain to be given a chance to polish a locomotive. But though the fireman occasionally let him hose down the boiler, he was never allowed to touch the paint or brass.

These, then, were the splendid machines that went hissing and sparkling into the world's first railroad war. When the Civil War broke out in the spring of 1861, nobody guessed that railroads would play much of a part in it. Even the voice of the industry, *The American Railroad Journal*, wrote that few of the lines would be affected by the struggle. But at the outset, everyone was making flawed predictions about the war, the chief one being that it wouldn't last very long.

Confederate General Robert E. Lee's superb lieutenant, Stonewall Jackson, early realized the importance of trains and managed to confiscate more than forty locomotives and three hundred cars. He and Lee shunted troops back and forth through the Shenandoah Valley on railroads with a speed unheard of in military history. Later, General Braxton Bragg sent his entire thirty-thousand-man Confederate army of the

Diesel locomotive, Penn-Central Railroad.

Tennessee from Tupelo, Mississippi, to Chattanooga—nearly eight hundred miles in a week. The Northern generals took longer to catch on; one, for instance, stopped his vital troop train overnight so his wife could catch up on her sleep in a nearby farmhouse.

But the North also had farsighted men who saw that the railroad was going to mean as much as any army corps or artillery battery to the outcome of the war. In 1862, Edwin Stanton, Lincoln's Secretary of War, appointed Daniel McCallum overall superintendent of military railroads in the United States. McCallum was a harsh, demanding, capable Scot who had found time, in between being superintendent of the Erie Railroad and inventing a new kind of bridge, to write popular poetry. The first man in history to have such a superintendency, he threw himself into the job with extraordinary vigor. At first his entire command consisted of seven miles of track in Virginia, but by the end of the war he ruled over seventeen thousand men, four hundred locomotives, and two thousand miles of road.

Stanton chose Hermann Haupt, another tireless railroad man, to work under McCallum in the East. He got hundreds of miles of track down, and trained men to build wooden railroad bridges in a matter of days. Lincoln, visiting one of these sturdy but spindly-looking structures, reported in wonderment, "That man Haupt has built a bridge across Potomac Creek, about four hundred feet long and nearly a hundred feet high, over which loaded trains are running every hour, and, upon my word . . . there is nothing in it but beanpoles and cornstalks."

Portage Viaduct over the Genesee River, Portageville, New York.

Once Haupt and McCallum got the Northern trains running efficiently, the South began to choke to death. To begin with, the nine thousand miles of Southern lines made up less than a third of the track in America. The Tredegar Iron Works in Richmond was the only factory in the South capable of turning out any significant amount of new track, but Tredegar had shifted to the manufacture of cannon at the outbreak of the war, and before long desperate Southern railroad men had to uproot whole lines in Florida and Texas in order to transplant them closer to the fighting. In the North, meanwhile, skies were stained with fire over a dozen steel towns, all of them feeding new rails to the Union — twenty-two thousand tons' worth by 1865.

When General Sherman began his Atlanta Campaign in 1864, McCallum had the trains running so smoothly that an average of a hundred and sixty cars a day rolled supplies forward to the Union fighting line. The Confederates retreating before Sherman's thrust pulled up track as they went, but Sherman's boys speedily replaced it: "The quicker you build the railroad," the general had told them, "the quicker you'll get something to eat." Even blasting shut a tunnel near Dalton, Georgia, did not slow down the Union troops. "Oh, hell," said one disgusted Rebel, "don't you know Sherman carries along a duplicate tunnel?"

Against all this energy and organization the Confederacy could bring only decaying equipment that took an appalling amount of money to keep running. In 1861 a gallon of lubricating oil cost a dollar in the South; three years later it cost fifty. Car wheels had

The Thebes Bridge, Mississippi River, Thebes, Illinois.

GT

risen from fifteen to five hundred dollars. Virtually all of the Confederate rail system lay in ruins. The booming North, on the other hand, had laid forty-five hundred miles of new track.

Still, the South kept its creaking railroads running, until on April 8, 1865, Union cavalrymen bushwhacked a Confederate supply train outside of a small Virginia town. The town was Appomattox, and the weary Southerners who might have gotten enough food and gunpowder to go on for a week or a month considered their well-supplied enemy and gave up.

The last train of the war was a Northern train. It rolled out of Washington, D.C., with a clear road ahead of it: "This train has the right of track over all other trains bound in either direction," the orders said. It was a short train and carried no troops or guns — and yet people who had long since grown tired of the endless military trains turned out by the hundreds to watch it pass by. They gathered thick around the city platforms when it went through, and came, two or three families to a wagon, to stand in the rain at country grade crossings, while their sleepy mules nodded, undisturbed by the sad, silver clang of the approaching locomotive. The people squinted down the nighttime track, saw the pale headlamp spill its light forward around the bend, and then watched the cars pass by, one of them carrying the murdered President home to Springfield, Illinois.

Three crowded years earlier, Abraham Lincoln had signed a Pacific Railroad Act which provided that track would be laid clear across the country. The act was still very much alive, and two companies were already at work on the road: the Central Pacific, which would lay track east from Sacramento, California, and the Union Pacific, which would move west from the Missouri River. The lines would meet on the California-Nevada boundary.

It is one thing to sign a bill authorizing a railroad, and quite another to build one. There was some railroad-building machinery around in the late 1860s, but for the most part the track would have to be laid by sweat and muscle and bare hands.

Between 1815 and 1860 two million Irish immigrants came to America. Doubtless many of them had come beguiled by the will-o'-the-wisp of streets of gold and an easy life; what they found were squalor and hard work and the Civil War. In the mud-marches and battles of the war, they became tough and disciplined. They would make an ideal labor force for the Union Pacific. Or so it seemed to Jack and Dan Casement as they looked over the hardy, bearded Irish veterans, still holding their service rifles, who had come out to earn two and a half dollars a day working on the railroad.

The Casement brothers had been hired as construction bosses by the Union Pacific, and a better choice could hardly have been made. Under the overall command of General Grenville Dodge — a sometime Northern railroad man who held supreme authority over the ten thousand men of the Union Pacific — the Casements worked out an assembly-line system of laying that was wholly new. People who saw it at work never forgot the sight.

At sunup the men would tumble out of their bunks in a twenty-car work train, a rolling city that also had room for feed stores, machine and carpentry shops, a company store, and a telegraph. A supply train would come nosing up behind the work train and unload rails, spikes, rods, fishplates, and all the other heavy paraphernalia that was necessary. This clutter was lifted onto flatcars and hauled out to the railhead — the farthest point of completed track — where the "iron men" were waiting. They would heft the seven-hundred-pound rails — five men to a rail — and at the order "Down!" drop them into place on the rail bed. The flatcar would move forward on the new tracks even while the spikes were being driven.

"It is a grand Anvil Chorus that these sturdy sledges are playing across the Plains," wrote an awed newspaperman. "It is in triple time, three strokes to a spike. There are ten spikes to a rail, four hundred rails to a mile, eighteen hundred miles to San Francisco — twenty-one million times are they to come down with their sharp punctuation before the great work of modern America is complete."

As the railroad crept west from Omaha, Nebraska, in a tumult of cursing and clanging, odd little rancid towns grew up around it. They would flourish for a few months, providing gamy recreation for the track crews, and then disappear. The process came to be known as "Hell on Wheels." Take, for instance, Julesburg, which sprang up like a boil from the empty desert 377 miles west of Omaha. In June it had a population of forty men and one woman; a month later four thousand people lived there and town lots were going for a thousand dollars each. The town was a riotous clutter of

Track gang laying rail, Penn-Central Railroad.

6177

saloons, whorehouses, and dance halls, and some called it "the Wickedest City in America." By September the railhead had been moved on and all that remained of Julesburg were heaps of rusting tin cans and a very full graveyard.

Hell on Wheels gave the men some diversion, but for the most part all they knew was work — as one writer later described it, " . . . the bare tracks pushing ahead across the bare prairie and the fussy cough of the wood-burning locomotives and the cold eyes of a murdered man, looking up at the prairie stars. And then there was the cholera and the malaria—and the strong man you'd worked beside, all of a sudden gripping his belly with the fear of death on his face and his shovel falling to the ground. Next day he would not be there and they'd scratch a name from the pay roll." The workers saw plenty of death, but they were hard men, and the Big Road, as they had begun to call it, roared along, two and three miles a day.

Miles away, the Central Pacific was chewing its way through the blind, snow-choked passes, of the Sierra Nevada, groping toward the Union Pacific railhead. In the passes, where drifts lay sixty feet deep, men chipped tunnels in granite so hard that the powder charges spurted out of blasting holes like fire from the mouth of a cannon, leaving the face of the rock undisturbed. A tough, indefatigable roughneck named Charles Crocker was in charge here, and most of his men were Chinese.

At first, nobody had thought the Chinese would be tough enough to stand the work in the Sierras, but Crocker, who didn't give a damn about anybody's opinion, had brought in fifty to see how they'd do. They had arrived quietly, set up a neat camp, went

to sleep with no roistering, and bounded up at sunrise to do a day's work that had Crocker wiring Sacramento demanding more workers as soon as possible. By the time he had bored his way through the mountains and had broken out into the Nevada desert, Crocker had more than six thousand Chinese building his road.

As the railheads approached each other, rivalry began to spring up between the crews. When Jack Casement's men put down six miles of track in a day, Crocker immediately beat his record by a mile. "No damned Chinamen can beat me laying rails," said Casement, and threw down eight miles. Crocker announced that he would lay ten, but when someone bet ten thousand dollars that he couldn't he appeared to back down. The next spring, however, with the road nearly finished, Crocker invited some important railroad officials to come out and see what a picked crew could do. He chose eight men, and gave them the signal to begin early in the morning. They worked straight through, with one short lunch break, until seven o'clock, When the last spike was driven, Crocker had his ten miles of track. His crew had handled 985,600 pounds of iron during the day, something over five and a half tons per man per hour. It is a feat that has never been equaled.

The rivalries took on harsher forms. Congress had made huge sums of money available for the project — up to $48,000 per mile in mountainous regions — and the Central Pacific and Union Pacific both wanted as much of it as they could get. So, when the two lines met, they actually passed each other and presented the weird spectacle of Chinese and Irish doggedly working away on two rights-of-way a few yards apart. The

Irish decided to settle things by setting charges of powder on the offending tracks and blowing the Chinese apart. Despite strong protests, the Union Pacific officials did nothing, and the grisly pranks continued until the Chinese quietly placed a charge of their own and buried several Irishmen. Thereafter, hostilities ceased.

Eventually the government interceded and, ordering the duplicate trackage abandoned, chose Promontory, Utah, as the place where the rails would be joined. On the cold, bright morning of May 10, 1869, two locomotives, one from each railroad, stood facing each other across a one-rail gap in the track. A band played while two crews — one Irish, one Chinese — laid the final rails. Someone waved to a nearby photographer and yelled, "Shoot." The Chinese, knowing just one meaning for that word, gaped at their Irish counterparts and fled. But order was restored, a golden spike was driven, the telegraph carried the news to every city in America, and the great task was done.

Afterward, everything reeked of scandal. Revelations of fraud and the pillaging of government funds came boiling up, and the Union Pacific went bankrupt. But this meant little enough to the men who had laid the steel. Their job done, many of them drifted away; others stayed on with the railroad for the rest of their lives. One of the latter is said to have remarked toward the end of his life, "There's an Irishman buried under every tie of that road." Today, freight trains still travel over parts of the old right-of-way, doing the business of the cities whose existence it made possible. There could be worse monuments to the anonymous men who left their youth and their bones on the Big Road.

The first Union Pacific train to Sacramento carried its share of dignitaries and, perched somewhere above or below the cars, an illegal passenger, one Omaha Bill. Omaha Bill, the first hobo to beat his way through to the Coast, was one of thousands of penniless men who went wherever the trains ran. Many of them had been shaken loose from their home towns by the war and found that afterward they never could quite get around to settling down again. Others were infected with that curious American restlessness which somehow makes the journey more satisfying than the goal. Others were simply looking for a job.

During the years after the Civil War hoboes became a terrifying legend, and all through the 1870s the papers reported tramp scares and published drawings of shaggy people who would tumble off a freight car, come into your kitchen, and butcher your family for a piece of pie.

In fact, tramps rarely did more harm than to pester housewives for a handout. Nor did all of them scour the country for free food and lodging. One tramp defined the three categories his fellows tended to fall into: the hobo, who worked and wandered; the tramp, who dreamed and wandered; and the bum, who drank and wandered. Hoboes, in fact, were as important to the building of the West as the railroads that carried them there. They would travel by the hundreds to distant communities where there was no real labor force, work for a week or two, and then move on to someplace else where a lot of men were needed for a little while. Winter would find them up on the Great Lakes,

Passenger station, St. Albans, Vermont.

ST. ALBANS
CENTRAL VERMONT RAILWAY INC.

bending to saws to bring in the ice harvest; then they might drift west to do a bit of mine work in Montana; then south to help gather the wheat of the endless Kansas prairies. Sometimes they put half a continent between one job and the next.

They kept their own company and developed their own slang, a rough, new kind of talk that, as a boy, the writer Jack London found intoxicating. "On the sand-bar above the railroad bridge we fell in with a bunch of boys likewise in swimming. . . . They were road kids, and with every word they uttered the lure of The Road laid hold of me more imperiously.

" 'When I was down in Alabama,' one kid would begin; or, another, 'Coming up on the C. & A. there ain't no steps to the "blinds." ' And I would lie silently in the sand and listen. 'It was at a little town in Ohio on the Lake Shore and Michigan Southern,' a kid would start; and another, 'Ever ride the Cannonball on the Wabash?'; and yet another, 'Nope, but I've been on the White Mail out of Chicago.' 'Talk about railroadin' — wait till you hit the Pennsylvania, four tracks, no water tanks, take water on the fly, that's goin' some.' "

London, enchanted, fell in with them and went on the bum. So did many other boys. Experienced hoboes often did their best to lure teen-agers away with them. The boys who went were expected to serve their mentors by foraging for food, cooking it, and — in that all but womanless society — providing more intimate services. In fact, the great anthem of all wanderers, "The Big Rock Candy Mountain," began its life as a

Currie, Nevada.

CURRIE
63

bitter, ironic song about a hobo telling extravagant lies in hopes of getting an innocent to join him on the road.

The song was written by Haywire Mac McClintock, a wonderful old adventurer who ran away from home at fourteen to join the circus and soon found himself bumming his way around the country in very tough company. "The decent hoboes were protective as long as they were around," he said, "but there were times when I fought like a wildcat or ran like a deer to preserve my independence and my virginity, and on one occasion I jumped into the darkness from a box-car door — from a train that must have been doing better than thirty miles an hour. I lay in the ditch where I landed until picked up by a section gang next morning."

It was with experiences like that in mind that McClintock first wrote his cynical ballad, but as it went the rounds from camp to camp where hoboes gathered, a wistful note came into the song, and in time it transformed itself into a utopian hymn to the land where the hoboes' road ended, "a land that's fair and bright, where the hand-outs grow on bushes," where there's "a lake of stew, and a gin lake too," and "there ain't no snow, where the rain don't fall and the wind don't blow, in the Big Rock Candy Mountain."

The hobo found this mythical land particularly appealing because his road was a rough one, marked all the way with hobo jungles where the men gathered around dim fires to eat mulligan stew — a vile amalgam of every kind of food they could scrounge — and water tanks where the trains stopped long enough for the non-paying passengers to scramble aboard the brakebeams where, once perched, a moment's inattention would send them to death underneath the iron wheels. And the road was well guarded by railroad bulls.

The bull was a private cop hired by the railroad to keep the hoboes away. The best of them became very good at their work indeed; for years hoboes stayed clear of the freight yards at Galesburg because of a particularly fierce bull, and Green River, Wyoming, was known throughout the country as the hunting ground of the terrible Green River Slim. Most feared of all was Jeff Carr of Cheyenne, Wyoming, who liked to discourage hoboes by shooting them dead.

The railroads wanted to be rid of hoboes not because they cheated the companies out of their fares, but because they caused a great deal of damage and stole vast amounts of freight. Nevertheless, the hoboes felt themselves persecuted. But men who owned the railroads were not sentimentalists, they were in it for the money. Most of them were ruthless, many of them outright buccaneers, and they looted more from the railroads than an army of hoboes could have carried off in a generation.

The first railroad owners were builders, but in the wake of the Civil War came a plague of flamboyant opportunists who saw that the fast money lay not in building lines, but in buying them and selling them for a profit. These men did their trading far from the dusty plains where workmen were laying iron, in offices along the nation's financial center of Wall Street. Foremost among these so-called robber barons was Cornelius

341

Vanderbilt, an unschooled fighter who had risen from doing odd jobs on the waterfront to own more steamships than any other man in the country. He was well into his sixties when railroads first caught his eye, but when he moved, it was with all of his strength. In a few years he had grabbed the Harlem, the Hudson, and the New York Central Railroad. That had been easy, but when he tried for New York State's other big railroad, the Erie, he came up against men as tough and devious as he. There were Jim Fisk, a steamboat man who liked to dress as an admiral; Daniel Drew, a canny old skinflint who dressed like a country deacon and whose greatest delight lay in bankrupting men who mistook him for a hick; and Jay Gould, perhaps the worst of all of them, a sallow, bearded, quiet man who had driven his first business partner to suicide and who had not looked back since.

The Erie War, as it came to be called, is a perfect example of how such men liked to operate, and is remarkable not so much for its total corruption as for the complete openness of that corruption. The newspapers of the day reported the dealings as they would a series of prize fights.

The war began in 1868 when Vanderbilt started buying up Erie stock. He thought he was making pretty good headway when all of a sudden a hundred thousand new shares came onto the market. Though momentarily baffled, Vanderbilt soon found the explanation; Fisk and Drew were printing phony stock certificates as fast as they could. "If this printing-press don't break down," said Fisk happily, "I'll be damned if I don't give the old hog all he wants of Erie."

Thurmond, West Virginia.

Whistle post on St. Johnsbury & Lamoille County Railroad.

(left) Freight train approaching El Paso, Illinois.

Of course, this was flagrantly illegal, and Vanderbilt prodded a New York judge into ordering the arrest of Fisk, Gould, and Drew. But they got wind of it and, bundling up the proceeds from their bogus stocks — some six million dollars in cash — they escaped across the Hudson River to Jersey City, where they holed up in Taylor's Hotel. They hired a hundred thugs to guard them there, and Fisk liked to call his stronghold Fort Taylor.

Gould, meanwhile, packed a suitcase with half a million dollars and headed up to Albany to buy some New York State legislators. He wanted them to make the stock issue legal. Vanderbilt, too, made his way north, and for a while there was a good deal of spirited bidding. Gould eventually spread a million dollars around; nobody knows how much Vanderbilt spent. The legislators upped their prices until a terrible rumor went around: Vanderbilt had withdrawn his opposition to the Erie bill. Prices tumbled, and politicians who had been holding out for five thousand dollars came to Gould offering to sell their votes for as little as a hundred. But it was too late.

Vanderbilt had sent a note to Drew, saying, "I'm sick of the whole damned business. Come and see me." They arranged a compromise, whereby Vanderbilt got some of his money back in return for calling off his lawyers. He never did get the Erie, though.

That was the only time Vanderbilt ever got his fingers burned in a business deal. At the end of his life he was not the richest man in the world — it was still the age of kings — but he certainly was the richest man in America. He left behind a widely publicized fortune of over a hundred million dollars when he died early in 1877.

That same year, the leaders of four major railroads — the Pennsylvania, the Erie, the New York Central, and the Baltimore and Ohio — decided it would be a good idea to cut trainmen's wages by ten percent. At that time a brakeman earned as little as a dollar for a twelve-hour workday, and the engineer, the highest paid trainman, got no more than three. Yet here were the railroads, many of them in good enough financial shape to pay generous dividends to their stockholders, cutting wages for the working man. It sounds like madness, and it was; it triggered off the first nationwide strike in our history, and the bloodiest.

On June 1, twenty thousand men on the Pennsylvania took the pay cut and kept on working. They were in a poisonous mood but, as the superintendent of the Pittsburgh Division explained, "The men were always complaining about something." Then, a month and a half later, the B & O's wage cut went into effect, and men walked off the job in Martinsburg, West Virginia. The militia turned out to suppress the strike but, confronted by an angry mob, suddenly found themselves in sympathy with the workers, shed their uniforms and went back home.

The strike burned its way along the B & O line. Railroad officials demanded that President Rutherford Hayes send the army to quell the "insurrection" and in a few days marines were patroling the streets in Baltimore. But within a week the strike spread

throughout the country. Of the seventy thousand miles of railroad track in the country, strikes shut down freight traffic on more than fifty thousand.

The worst violence came in Pittsburgh, where a thousand troops arrived to force the men back to work at bayonet-point. Crowds stoned the troops as they marched into the city, and the soldiers opened fire, killing twenty men, a woman and three children. A furious mob of fifteen thousand drove the soldiers into a roundhouse, then set it afire. The troops finally fought their way out while the flames spread to the boxcars in the freight yards, then to the city. When firemen came to fight the blaze, one group of rioters trained a cannon at their hose wagon, shouting, "If you don't get out of that, we'll blow you to hell." Pittsburgh burned unchecked all night long.

In Buffalo and Chicago, in Boston and Omaha, mobs fought the railroad. Officers who had served in the Civil War fifteen years before found themselves once again facing off against their countrymen.

In two weeks the worst of the violence had burned itself out. A hundred people had been killed, five hundred injured, millions of dollars' worth of railroad property destroyed. In Pittsburgh alone one hundred and four locomotives and more than two thousand cars had been gutted.

Some lines restored the old wages, but for the most part the strike had been broken. Nevertheless, things would not be quite the same again. The workers had learned their strength; later they would learn how to use it, and in time would form the

Sal Bagnoli, yard conductor, Delaware & Hudson Railroad.

railroad brotherhoods, some of the most effective labor unions in the country. And, too, the financiers had come under a shadow. One Chicago newspaperman, after voicing the obligatory condemnation of the rioters, went on to say that the public had even less sympathy for "the Vanderbilts, the Jay Goulds, and the Jim Fisks who have been running the railroads and have ruined one of the finest properties the world has known." Powerful men would still buy and sell, inflate and ruin railroads at their whim, but never again with quite so free a hand as they used during the gaudy, untrammelled days of the Erie War. From now on, they knew, they were being watched.

While trainmen were warring against soldiers for a living wage, a strange, scrawny, bearded fanatic named Lorenzo Coffin was fighting a very different battle on their behalf. Coffin, an Iowa farmer, took up the trainmen's cause on a day in 1874 when, riding a train, he idly watched a brakeman helping to couple together two boxcars. The couplings in those days were called link-and-pin, evil devices that required the brakeman to guide two iron loops together while standing between the cars and then fasten them by dropping a bolt through. As Coffin watched the cars meet, he saw the brakeman go down, screaming, with the two remaining fingers on his right hand torn away; he had lost the other two the same way the year before.

Horrified, Coffin began to investigate the accident and found that it was all in a day's work for brakemen. An experienced man with a full set of fingers was a rarity; in fact, employers tended to look for missing fingers when choosing a trainman for a job. By and

Brakeman coupling cars, Whitehall, New York.

Fred J. Lewis, freight conductor, Delaware & Hudson Railroad.

Setting hand brake.

Coupling air brake hose.

large, Coffin discovered, the life of a brakeman was just about as dangerous as that of a gunfighter. Brakes had to be hand set from the tops of cars, and a bad winter run would find men slipping and teetering back and forth along the train, straining at icy brake wheels. Hundreds of men died every year, and thousands more were maimed.

Coffin pressed on. Was there no alternative to the butchery? Yes, there was. George Westinghouse had invented a brake that locked the wheels through compressed air, and a man named Eli Janney had developed an efficient automatic coupler that clamped like two hands shaking when the cars came together. Why weren't these devices being used? Because, for their dollar-fifty a day, brakemen were held responsible for their own injuries. Air brakes and automatic couplers cost money; brakemen cost next to nothing.

This set Coffin off on a crusade that lasted twenty years. He became a familiar and well-hated figure in railroad offices, where he went to badger the officials. Eventually they simply had "the air brake fanatic" thrown out. Determined, Coffin wrote hundreds of letters, pressured everyone connected with the railroads and, in 1887, finally provoked his opponents into staging a test. It was held on a steep grade on the Chicago, Burlington & Quincy right-of-way, and Westinghouse came to watch the results with Coffin. A long freight started to roll downgrade, and when it reached forty miles an hour, the engineer threw on the brakes. The train came to a dead stop within five hundred feet. Coffin, weeping, said, "I am the happiest man in all Creation!"

Still the railroads fought Coffin. The tireless monomaniac lobbied year after year for a nationwide law. At last, in 1893, he triumphed; President Benjamin Harrison signed the Railway Safety Appliance Act into law, and from that time hence airbrakes and automatic couplers were mandatory on all trains. To his huge satisfaction, Coffin saw railroad employee accidents decline 60 percent almost immediately. And after a while even the railroad companies saw that better safety records meant better revenues, and that the new devices allowed heavier, faster trains than had been possible in the murderous link-and-pin days.

So the railroads came clanking into the twentieth century and, indeed, they had hauled the nation along with them. The whole character of American civilization had changed since that July Fourth when old Charles Carroll turned his spadeful of dirt. No longer were we a country of isolated, self-sufficient communities; the nation was interconnected from east to west and from north to south, and cities had begun to act as giant specialty shops, each supplying a product to the rest of the country. Chicago dressed its meat, St. Louis brewed its beer, Grand Rapids made its cheap, sturdy furniture, Minneapolis milled its grain. These products, and thousands of others, all moved by rail. Getting the freight cars where they were going, and keeping track of them along the way, required a terrifying amount of organization. There usually wasn't a direct train from, say, Chicago to Portland, Oregon. A carload of meat might be hauled first from Chicago to San Francisco, then be coupled to another train going north to

Seattle, and finally get hooked to a local freight that would pull it to its destination.

The fast freight clattering along through the countryside—that was the easy part of the job. The hard work of railroading took place in the freight yards outside every fair-sized town, a baffling crisscross of closely laid tracks cluttered with engine houses, water tanks, icing stations, repair shops, and signal towers.

The switching yards outside of the great depot of Chicago grew to cover an area larger than the entire state of Rhode Island with some eight thousand miles of track. To this monster yard, and to all the smaller ones, came freight cars from all over, bearing on their sides the names of the lines that owned them: Santa Fe, Burlington, Nickel Plate, Erie, Wabash, Great Northern. In the yard, the trains went first onto a receiving track, off the main right-of-way, and then into the classification yard, where they were broken apart, each car checked and shunted aside to become part of another train. Night and day, hardy, blunt-nosed little switch engines shuffled back and forth, pushing and pulling freight cars.

The laborious process became a little smoother when the lines developed hump switching, a way to let gravity, rather than engines, do the bulk of the work. An engine would push a string of cars up a hill, and, as they passed the crest, they would be cut loose from the train either singly or in small groups. Controlled by brakemen, the cars would roll down the hill until they reached their designated tracks. There, a switchman would throw a switch and the cars would coast into their slots.

Engine terminal and freight yard, Bluefield, West Virginia.

8522
8522
NW

(left)
Freight train arriving in classification yard, Oneonta, New York.

"Pulling the pin," Mechanicville, New York.

Freight car wheels, Juniata Shops, Altoona, Pennsylvania.

(right) Repairing boxcar coupler, Penn-Central Railroad

At first, the humpmaster had the maddening job of figuring out a car's classification, as it came over the hill, from chalk marks scrawled on its side. These had been put on by the yard clerks who met the train as it came in; and they made their marks according to waybills handed them by the conductor. Eventually, telegraph-typewriters came to be installed, and they rattled out the information long before the train reached the yards. Soon, too, a system of automatic controls, monitored from a single point, replaced the individual switch tenders. The process has been speeded up over the years and modern automated classification yards can handle thousands of cars a day.

Some cars were unloaded in the yards and, after 1900, the law said these empties could only be sent back in the direction of the railroad that owned them. The line that owned the car charged the line that had rented it, at first by the miles traveled, and later by the day. Though the rental was no more than a few cents daily at the turn of the century, with thousands of cars scattered throughout the nation it could mount up to real money. All over the country, harried men had to keep track of these wanderers. A freight car from Maine's Bangor & Aroostook Railroad might be standing idle in a San Francisco yard, looking as lost and derelict as a beached ship, but somewhere somebody had entered its number in a ledger, and knew—or was supposed to know—just where it was.

Hump yard, North Platte, Nebraska.

NP

Boxcar, Canadian National Railways.

Boxcar, Napierville Junction Railroad.

Covered hopper car, Pennsylvania Railroad.

Caboose, Soo Line Railroad.

Sometimes, cars disappeared for months. On December 14, 1886, a loaded car rolled out of its home yard in Indianapolis headed for Boston. It got there, after a lot of delays, on New Year's Eve. But then another railroad hauled it off to Chicago, and then to St. Paul, where it was reloaded and sent to New York. It then looped up to Canada, while its owners began to demand its return. Back it went to Chicago, and then straight on to Newport News, Virginia; it had passed through Indianapolis on its way, but the angry owners didn't find out about that until later. While they screamed, the car went down into South Carolina and Georgia, then out to Selma, Alabama. Finally, having generated dozens of threatening and pleading telegraph messages, the weatherbeaten veteran creaked into the Indianapolis yards on April 17, 1888, having been away from home for sixteen months.

That misrouting had been half accident, but on the smaller lines empty cars stood a real threat of being stolen. One old trainman remembered the technique: "Whenever we could get hold of an empty boxcar," he said, "we generally took it. . . . I recollect one day the Kansas City, St. Joseph & Council Bluffs yardmaster put eleven cars on a sidetrack near the east end of the bridge, while he went to dinner, after which he was going to take them to the freight house to load with freight for his road. While he was gone, I rushed an engine over the river and pulled those eleven boxcars to Elwood and hid them in the woods, so we could have cars when we needed them. The only way we could move what

Flat switching yard, Oneonta, New York.

little business we had to move was by hooking somebody's equipment to do it with."

Barring theft, carelessness, and accident, when the classification process was finished, the cars ended up in the yards strung together into new trains. Once they were assembled, it remained only for a caboose to be coupled onto the end, the air brakes tested, and a road engine, which was longer and taller than the switch engines, to be hooked onto the front. Then the conductor would get new waybills from the yard clerk, and the engineer would run the train out of the yard onto the main line.

The engineer had played little part in the classification process; in fact, road engineers were not even paid for any switching time they might have to put in. Nevertheless the engineer was the essence of railroading, and therefore the essence of glory, to thousands of small boys whose only desire in life was to take his place at the throttle. There is simply no modern equivalent to the supremely powerful hold the railroad engineer held over the popular imagination for three quarters of a century. His job meant adventure, speed, and the thrilling chance to see what lay beyond the dull, predictable, eternal routines of farm life. There are still people alive today who will tell you that one of the truly great moments of their lives was standing by the water tank in a lost prairie town and having the engineer grin down from his cab to say something like "Hey, kid, you can ride up here with me as far as Haddensburg, if you don't mind the walk back."

Locomotive engineer, Delaware & Hudson Railroad.

It is a fair gauge of his romantic power that the only railroad man everybody can name today was an engineer. He was a good one, but not remarkable, and the run that made him famous started out as a perfectly average one. But he made it in 1900, the high noon of American railroading, when there were two hundred thousand miles of main-line track in the country, and everyone thought the railroad train would hold its position of power and prominence forever.

This engineer's given name was John Luther Jones, but he hailed from Cayce, Kentucky, so his friends took to calling him Casey. Casey Jones had three railroading brothers; in those days, chances were that if one member of a family worked on a railroad, others did too. He began his career as a lightning-slinger, one of the thousands of men who worked as railroad telegraphers. In 1888 he signed on as a fireman on the Illinois Central, whose vast network of track stretched from the Great Lakes to the Gulf of Mexico. After two years he graduated from the left- to the right-hand side of the cab and took his place at the throttle as a full-fledged engineer. Soon he was pulling fast freights in and out of Centralia, Illinois. His fireman in those days remembered the young engineer as a "long, lean, lanky man who was so tall that he couldn't stand up in the engine cab without sticking his head outside a foot or so, reminding some of his friends of a young giraffe."

Casey had a touch of recklessness to him, and he drew nine reprimands for breaking the iron-bound railroad rules, each time getting suspended from five to thirty days. But he stayed clear of liquor, nobody had ever died as a result of his carelessness, and he had

Train orders, telegraph office, East Salamanca, New York.

a reputation as a man who could get a train in on time. And so, in his twelfth year with the I.C., he was offered one of the best jobs on the line: engineer on the fastest Chicago-New Orleans passenger train, No. 1, the *New Orleans Special*. The men called it the *Cannonball;* railroaders always liked to call a crack train the *Cannonball* or the *Flyer*. Casey was to haul it on the one-hundred-ninety-mile run between Memphis and Canton.

The trip that made Casey immortal began on the evening of April 29 after he had already done a full day's work. He and his fireman, Sim Webb, had already brought the northbound *Cannonball* into the Poplar Street Station in Memphis behind engine No. 382. But when they took their locomotive into the roundhouse they found that the engineer scheduled to drive the *Cannonball* on her southbound run had called in sick. So it fell to Casey to go back over the two-hundred-mile route he had just traveled. He said he'd do it, so long as No. 382 got a thorough going-over in the roundhouse. Casey and Sim got a little food and rest as roundhouse mechanics checked the engine while the huge exhaust fans roared overhead, and other locomotives snorted in and out of the dark, smoky building.

Casey and Sim showed up ready to leave the terminal on schedule at 11:15 P.M., but there was no sign of the *Cannonball*. Casey walked up and down the tall flanks of his locomotive, performing the engineer's old ritual of poking into the gleaming running gear with his oil can. At last, when he'd done all that needed doing, he returned to the

"Hooping up" train orders to westbound freight leaving Oneonta.

FA

cab, took hold of the grab-iron, and pulled himself up onto the deck. There, he and Sim fretted until they heard a whistle squalling in the distance. The *Cannonball* was coming in at last, ninety minutes late.

Cracking the throttle, Casey drifted down toward the switch while the other engine uncoupled and headed for the roundhouse. Casey rolled out ahead and hooked onto his six-car drag of coaches, baggage, and mail.

He stopped briefly at the station, and his conductor came up and handed him a sheaf of flimsies. Casey thumbed through these train orders and saw that he owned the railroad; the track would be cleared for him all the way through to Canton. He was to make up time on the way, getting into Grenada, a hundred miles away, thirty-five minutes late, into Durant, fifty-three miles beyond, twenty minutes late, and into Canton on time. He turned to Sim. "We're going to have a pretty tough time getting into Canton on the dot," he said, "but I believe we can do it, barring accidents."

He pulled out of Memphis ninety-six minutes off the advertised schedule, opening the throttle as he went through the yards, shrilling on his whistle. It was distinctive, that whistle; it had six different pipes, and Casey could just about play a tune on it. The night was close and murky, and clouds hung low over the train as Casey opened her up, doing better than seventy on the long, straight stretch to Grenada. And all over the country that night, trains were moving through the darkness — fast freights loping along the New York Central's fine four-track right-of-way, tiny, toylike logging narrow-gauges fussing and chirping through tall timber, big tenwheeler locomotives chopping their way upgrade through the Sierras, where the snow still lay deep in the high passes.

No. 382 was running beautifully and Casey roared along while Sim fed coal into the firebox. They made Grenada thirty-five minutes behind schedule, as ordered. "We were whittling that lost time away to nothing," said Sim, "and Mr. Casey was still in high spirits. As we left Durant, he stood up and hollered to me over the boiler head: 'Oh, Sim! The old girl's got her high-heeled slippers on tonight! ' "

Down the line at Vaughan, twelve miles above Canton, an ugly situation was shaping up. A southbound freight had gone into the hole—that is, run onto a passing track — to let the *Cannonball* go by. A northbound freight came in shortly after. Together the two trains were four cars too long for the passing track, and the southbound had to edge back out onto the main line to let two passenger locals get onto a siding. But before the southbound freight could get clear of the track again, the northbound train facing it blew an air hose, locking its brakes. It couldn't move, and so four freight cars were left standing on the main line.

Cursing men ran forward and placed a warning torpedo—a small explosive charge that made a sharp noise when train wheels hit it—on the track. Soon No. 382's headlight came around the bend. The tall locomotive screamed past; the torpedo cracked, a flagman frantically waved his red lantern.

Up in the cab, Sim squinted through the drizzle and saw the lights of a caboose dead ahead. "Look out!" he yelled. "We're going to hit something!"

"Jump, Sim!" shouted Casey, as he kicked his seat out from under him and stood up to grab the brakes. Sim leaped into the night, while Casey dynamited the train; he released sand onto the rails, pulled on the brakes, and swung the Johnson bar over so that the engine went into reverse. Sparks sprayed out from the shrieking wheels; the engine shuddered and slowed, and Casey stayed with her, fighting for every inch. He had cut his speed in half by the time he slammed into the freight train. The caboose virtually exploded into flailing timbers; No. 382, with all the pipes sheared off her, jumped the track, swung halfway around and, wheels spinning, plunged into a ditch where she lay bleeding steam and oil.

They found Casey in the wreckage. He still had one hand on the airbrakes they said, and one on the whistle cord. He had kept his safety record intact; he alone died in the wreck.

That was Casey Jones's trip to the promised land. There were lots of wrecks like it that year—lots of wrecks like it every year. Many other railroad men's wives went wild with grief that season when they got the news, just as Jane Brady Jones did when the grim trainmen showed up to tell her about Casey. And Casey would have been forgotten today had it not been for a black engine wiper named Wallace Saunders, who worked in the Canton roundhouse. Saunders heard all about the crash up at Vaughan, of course, and as he worked, he began to put together a song. The song went the rounds of the I.C., and eventually got picked up by a professional hit-smith who fashioned it into the inanely jaunty ballad that everybody knows today.

But Haywire Mac McClintock, he of the Big Rock Candy Mountain, drifted through the Canton yards shortly after Saunders made up the song, and got the old roundhouse man to sing it to him. Years later, he sang Saunders' original version, and it is a sad, haunting lament. The skeleton of the popular tune is there, but none of the bounce; it is, rather, a blues, a dirge for all the men who lost an arm or a leg or a life on the railroad:

> You oughta been there, for to see the sight,
> All the women were crying, both colored and white,
> I was there, for to tell the fact,
> They flagged him down, but he never looked back.

As for No. 382, they rebuilt her and put her back pulling the *Cannonball.* But she never behaved as well after the wreck as she had before, and the engineers came to regard her as a jinx. In 1903 the engine turned over at a switch, killed a fireman, and badly injured an engineer. During the next few years she jumped the tracks no less than four times, and finally, in 1935, on her way to be scrapped, she made one last lunge and killed another fireman.

A lot happened during the thirty-five years that passed between the time Casey took No. 382 out of Memphis and the time she went unlamented to the scrap heap. Steam locomotives grew larger than anything Casey had ever seen, or could have imagined—monsters with twenty driving wheels that looked as long as a city block and could pull one hundred-and-sixty-car trains all by themselves. But even as steam locomotives grew to their final, greatest size, they were doomed.

(right) Engine terminal, White River Junction, Vermont.

Steam locomotive.

Steam locomotive taking water,
Vallée Jonction, Quebec.

Washing crosshead of steam locomotive, McAdam, New Brunswick.

Driving wheels, steam locomotive, Canadian Pacific Railway.

(right)
Steam locomotive
leaving roundhouse.

1262

In 1925 a strange, square little switch engine appeared in the yards of the Central Railroad of New Jersey. It was the first diesel locomotive in America, and the engineers gliding through the yards in their high-wheeled steam engines must have thought it was quite a joke. But within half a century, it would knock every single steam engineer out of his cab.

A diesel locomotive has an internal combustion engine that burns oil. The engine does not turn the wheels directly, but runs a generator which produces electric power that drives the train. In a few years the diesel had evolved into a rugged competitor to steam. The engines spent less time undergoing repairs, and they were much, much easier to get going. Moreover, they were flexible; a big steam engine produced the same amount of horsepower whether it was pulling ten cars or a hundred, while diesels could be paired up or used singly as the load they had to haul demanded.

And so the diesel began to edge out steam, and eventually replaced it entirely. The last commercially built steam engine in America rolled out of the Lima locomotive works on Friday, the thirteenth of April, 1949.

Though it was sad to see the age of steam drawing to a close, American railroads were coming into a far more desperate era. During the 1950s and 1960s the great system which had helped build the country began to disintegrate. First the automobile and then the airplane chipped away at freight and passenger travel. And, too, the railroads began to pay for the sins of their fathers. The legacy of hatred and mistrust built up by the

Fisks and the Goulds eventually resulted in stiff laws, laws which are still punishing the railroads. No trucking company could last long if it had to pay full taxes, maintenance, and construction costs on the roads the trucks travel, but that is exactly what railroads must do.

Thanks to automatic classification yards, many lines still stay prosperous by hauling freight. But many more have gone bankrupt, and passenger service has dwindled until today railroads carry less than one percent of national passenger travel in the country. Switchyards lie all but empty with a few freight cars tucked away on a track as though they are going to stay there until they fall apart. Depots, which used to be the hub of small-town life, stand deserted; you see one every now and then, listing in a sea of weeds, as you speed past on the highway.

And yet, the railroads may spring back into life. We have just begun to realize what a thirsty, expensive luxury the automobile really is. As gasoline begins to approach the price of perfume, people are looking again to the trains, and what they see is one of the most efficient forms of transportation ever developed. A railroad can carry fifty thousand people an hour over a single track; to move that many over a highway, you need four lanes and ten thousand cars, all spewing their poison into the atmosphere. Railroads hardly pollute at all.

So the American railroad may return in its old power. But whether it does or not, it can never really leave us, because we are Americans, and trains are in our blood.

Ticket office window, Kingston, Rhode Island.

Depot, Martin,
North Dakota.

My father, the man who used to take me down to see the old steam engines going by on the Putnam Division, lives fifteen miles outside of New York City in a town on what used to be the New York Central, and what is now part of a big amalgam of different railroads known as Conrail. Last winter I took the train out to visit him on a gray Sunday. Toward dusk I said good-bye and went down to the station to go back to New York. It had started to snow at about two o'clock, and by four, when I stood looking out through the dingy windows of the station waiting room, a full blizzard was coming down.

All the trains were running late, and people were packed into the small, muggy room. A little kid sat next to me. He was about five years old, and he kept whining and pouting and saying, "When can we go home? Daddy, I want to go home." His daddy gave him angry pats on the head and said, "As soon as the train comes."

"I hate this," said the boy. After a while, he threw up, and I went out on the platform to wait in the storm. Soon I heard a long, low whistle, and a light showed, way up the track. I thought it was the New York local, and so did everybody else, because they all came up on the platform, the father and the child with them. Then the train came. It was an express, down from upstate, pulled by a big diesel, and not about to stop.

Now, for my money, a diesel looks like a loaf of bread compared with a steam engine. But this one came around the bend through the storm with the snow spiraling off its headlight and the engineer blowing the whistle to show he was going through, and it all looked just great. The people on the platform jumped back, but the little boy broke

Union Pacific freight train climbing Sherman Hill in Wyoming.

away from his father and ran forward to the edge of the tracks and jumped up and down, waving. The engineer saw him and gave him a special toot, and the train disappeared down the tracks into the blizzard, leaving the arcing noise of the whistle shaking on the snow behind it.

The boy kept dancing on the platform long after the train had gone, and I understood what had gotten him excited.

A man named Glen Mullin, who bummed around in between jobs, and who liked to call himself the "scholar-tramp," watched a lot of trains go by. Here's what he thought of them, and it's what most of us still think of them, despite the buses and the trucks and the airplanes:

"A train is a thing compounded of magic and beauty. . . . The rattle and swank of a long freight pulling out of the yards, the locomotive, black and eager, shoving her snorting muzzle along the rails, this is a spectacle and a challenge. . . . She is an enchanted caravan moving into the mysterious beyond, hailing with bells and song the blue distance that fades forever as she moves."